OWN YOUR AUTHORITY

Why Every Real Estate Professional Needs Their Own Website

authority
/ə'THôrədē/

the power to influence others, especially because of one's commanding manner or one's recognized knowledge about something.

lifestyle
/'līf,stīl/

the way in which a person or group lives.

equity
/'ekwədē/

digital assets that appreciate over time

Marna Friedman

OWN YOUR AUTHORITY:

Why Every Real Estate Professional Needs Their Own Website

ISBN 979-8-9893165-5-7
Book and cover layout and design
by Marna Friedman
Published by The Organized Agent, Inc.
www.theorganizedagent.com

Links to Other Websites and Services

Own Your Authority provides links to websites in our book for informational
purposes only. We do not assume any responsibility for those sites and
provide the links solely for the convenience of our readers. We neither own or
control the content of these sites and take no responsibility for their content,
nor should it be implied that we endorse or otherwise recommend such sites
or the products or services offered.

DISCLAIMER:

Every effort has been made to ensure the accuracy of the information provided.
However, we cannot be held liable for any errors, omissions, or inconsistencies.
The strategies and information provided are for informational purposes only and do
not constitute legal, financial, or professional advice. You are encouraged to consult
with qualified legal, financial, or other professional advisors before making any decisions or taking any actions based on this content. All real estate professionals should
always remember to abide by all industry rules related to their business including but
not limited to the Fair Housing Act, RESPA, REALTOR® Code of Ethics, brokerage
compliance and local/state board guidelines.

The reader acknowledges that they assume full responsibility for the use of the
materials and information presented in this publication. Compliance with all relevant
laws and regulations, including advertising and conducting business in the United
States or any other jurisdiction, lies solely with the reader. The author and publisher
bear no responsibility or liability for any reader's use of these materials.

Table of Contents

The Real Estate Authority Crisis No One Is Talking About

Every day, thousands of competent, experienced real estate professionals lose potential clients to agents they've never heard of. Not because these competitors are better negotiators, have superior market knowledge, or provide exceptional service. They lose clients because when consumers search for real estate expertise online, these capable professionals are invisible.

According to the National Association of Realtors, there are over 2 million active real estate licensees in the United States. Yet when potential clients research "*best realtor in [city]*" or seek answers to questions like "*Is now a good time to buy?*" or "*What should I know about this neighborhood?*"—the vast majority of these professionals simply don't appear in the results.

This isn't a story about incompetent real estate professionals failing to succeed. This is about competent professionals who have unknowingly made themselves invisible in the digital marketplace where 97% of consumers now begin their real estate journey.

Why This Book Exists

I've spent years helping businesses establish their digital presence and build authority in their markets. As a business consultant who has guided numerous companies through website development and digital marketing strategies, I've witnessed a troubling pattern in the real estate industry: exceptional professionals who possess deep market knowledge and provide outstanding service, yet remain completely invisible to the consumers who need their expertise most.

The problem isn't lack of knowledge or poor service—it's a fundamental misunderstanding of how authority and trust develop in the digital age. Most real estate professionals have built their entire online presence on platforms they don't control: broker websites, social media profiles, and portal listings. They've unknowingly chosen to rent their digital identity rather than own it.

This book exists to change that dynamic.

What's Different About This Approach

This isn't another real estate marketing book filled with outdated tactics about keyword stuffing and cold calling. Nor is it a generic guide to social media posting and lead generation. This book addresses the deeper strategic challenge facing real estate professionals: how to build genuine digital authority that commands respect, attracts quality clients, and creates sustainable competitive advantages.

THREE KEY DIFFERENCES SET THIS APPROACH APART:

1. **Authority Over Activity**

    While most real estate marketing focuses on generating activity—more leads, more social media followers, more advertising reach—this book   focuses on building authority. Authority attracts quality clients who seek your expertise specifically, rather than prospects who chose you based on convenience or price.

2. **Ownership Over Renting**

    Instead of building your professional reputation on platforms controlled by others, you'll learn to create digital assets you actually own. Your website becomes digital real estate that appreciates in value, building equity in your personal brand regardless of brokerage changes or platform algorithm shifts.

3. **AI-Era Optimization**
   This book addresses the fundamental changes in how consumers discover and evaluate expertise in the age of artificial intelligence. You'll understand AI-GEO (AI-Generated Experience Optimization), LEO (Local Experience Optimization), and AI-LEO (Language Experience Optimization)—concepts that most real estate professionals haven't even heard of yet.

Who This Book Is For

This book is written specifically for real estate professionals who recognize that the digital landscape has fundamentally changed, but aren't sure how to adapt effectively. You might be:

The Experienced Professional who has built a successful practice through referrals and networking, but realizes that newer agents with strong digital presence are capturing market share from prospects who research online before making contact.

The Mid-Level Producer who generates consistent business but wants to break through to the next level of success and recognition in your market. You understand that digital authority could be the differentiator that elevates your practice.

The Forward-Thinking Agent who recognizes that artificial intelligence and evolving search patterns are changing how consumers discover and evaluate real estate expertise. You want to position yourself ahead of these changes rather than react to them after competitors have already adapted.

The Professional Seeking Independence who wants to build genuine business equity that transfers with you regardless of brokerage affiliations, market conditions, or industry changes.

What You Won't Find in This Book

Before we begin, let me be clear about what this book does not contain:

Generic Marketing Tactics:

This isn't a collection of social media tips, email templates, or advertising strategies that apply to any business. Every recommendation is specifically tailored to real estate professionals and the unique challenges of building authority in local markets.

Quick-Fix Solutions:

Building digital authority requires time, consistency, and strategic thinking. If you're looking for tactics that promise immediate results with minimal effort, this book will disappoint you.

Platform-Dependent Strategies:

While we'll discuss how to leverage various platforms effectively, the core strategy focuses on building assets you control rather than depending on social media algorithms or portal policies that can change without notice.

Technical Implementation Details:

This first book establishes the strategic foundation and explains why website ownership is essential for real estate professionals. The detailed technical guidance for building and optimizing your website comprises the second book in this series.

How This Book Is Organized

Own Your Authority is structured to build understanding progressively, from identifying the problem to developing comprehensive solutions:

Chapters 1-2 establish the foundation by explaining the authority crisis facing real estate professionals and why website ownership has become a business necessity rather than a marketing option.

Chapters 3-4 explore how to build authority through content and understand the psychology of trust in high-stakes real estate transactions.

Chapters 5-6 provide frameworks for converting authority into business results through effective lead capture and multi-platform strategy integration.

Chapters 7-8 address advanced topics including AI-era optimization strategies and integration with existing real estate tools and broker resources.

Each chapter builds on previous concepts while providing actionable insights you can begin implementing immediately. However, the full benefit comes from understanding the complete strategic framework rather than implementing isolated tactics.

A Personal Note About Implementation

Having worked with hundreds of businesses on digital strategy implementation, I've learned that success comes not from perfect execution, but from consistent action guided by clear strategic understanding. The real estate professionals who achieve the best results from these concepts aren't necessarily the most technically sophisticated—they're the ones who understand the strategic importance of digital authority and commit to building it systematically over time.

You don't need to be a marketing expert or technology specialist to succeed with these strategies. You need to be willing to think strategically about your professional development and commit to building genuine value for your market rather than just promoting your services.

The Opportunity Ahead

The real estate industry is experiencing a fundamental shift in how consumers discover, evaluate, and select professionals. Artificial intelligence is changing search patterns, younger demographics are entering the market with different research habits, and traditional authority signals like years of experience or brokerage affiliation carry less weight than demonstrated expertise and digital credibility.

This shift creates unprecedented opportunity for real estate professionals willing to adapt their approach. The agents who build comprehensive digital authority now will have sustainable competitive advantages that strengthen over time, while those who continue depending on traditional marketing methods will find themselves increasingly invisible in the marketplace where consumers make their initial decisions.

The choice is simple: Continue building your professional reputation on platforms you don't control, hoping that current marketing methods remain effective as the industry evolves. Or begin building digital authority assets that you own, that appreciate in value over time, and that position you as the recognized expert consumers seek when making one of the largest financial decisions of their lives.

This book will show you exactly how to make that transformation.

The strategies, frameworks, and insights that follow represent years of business consulting experience combined with deep analysis of how digital authority actually develops in local service markets. More importantly, they're designed specifically for real estate professionals who understand that their expertise deserves to be discovered by the consumers who need it most.

Your knowledge and experience have tremendous value. The question is whether that value will remain hidden on platforms you don't control, or whether you'll build the digital authority necessary to ensure your expertise reaches the clients who need it most.

Let's begin building your authority platform.

Chapter 1

The Authority Crisis in Real Estate

The Invisible Real Estate Agent Problem

According to the National Association of Real Estate Professionals, there are over 2 million active real estate licensees in the United States. Yet when consumers search for local real estate expertise online, the vast majority of these professionals remain completely invisible. They've built their entire digital presence on platforms they don't control—broker websites, portal profiles, and social media accounts—without realizing they're competing in a marketplace designed to commoditize their expertise.

This invisibility isn't just a marketing problem. It's an authority crisis that affects income, client quality, and long-term business sustainability.

The Commoditization Trap

The real estate industry has systematically trained consumers to view real estate professionals as interchangeable service providers rather than trusted advisors. Consider what happens when someone begins their home search:

The Typical Consumer Journey:
1. They browse their preferred online portal
2. They see properties, not the listing agents
3. When they do see agent information, it's limited to a small photo and phone number
4. Every agent looks identical: professional headshot, 5-star reviews, "I'll work hard for you" tagline

In this environment, consumers make decisions based on convenience, commission rates, or random recommendations rather than expertise and authority. The platforms have successfully made the property the star while reducing real estate professionals to background players.

The Broker Website Illusion

Most real estate professionals believe their broker's website solves their online presence needs. After all, it has:

- Professional design
- MLS integration
- Lead capture forms
- SEO optimization
- Their bio and contact information

But here's what they don't realize:

> *You're not building authority-*
> *you're renting space in someone else's authority.*

When clients visit your broker's website, they're engaging with the brokerage brand, not your personal brand. The content, the expertise, the trust signals—they all point to the company, not to you as an individual expert.

The Three Pillars of Digital Authority

True authority in real estate—the kind that commands premium fees and generates consistent referrals—rests on three pillars that broker websites cannot provide:

1. **Owned Expertise Platform**
   Your knowledge and insights need a stage where YOU are the star. When you consistently publish market analyses, neighborhood insights, and buyer/seller education content,

you position yourself as the local expert. But this only works when the platform amplifies YOUR name, not your broker's.

2. **Personal Brand Equity**

Every piece of content, every interaction, every valuable resource should build equity in your personal brand. When you change brokers (and statistics show most real estate professionals change brokers multiple times), what happens to all that digital equity? If it's built on your broker's platform, you lose it all and start over.

Direct Relationship Building

The most successful real estate professionals don't just generate leads—they build relationships with their market over time. This requires consistent, valuable communication that positions them as the trusted advisor before, during, and after transactions.

The Content Authority Model

The future belongs to real estate professionals who understand that authority comes from consistently demonstrating expertise, not just claiming it. This means shifting from the traditional real estate marketing model to what I call the Content Authority Model.

Traditional Model:

- "Call me, I'm the best"
- Generic marketing materials
- Reactive lead response
- Transaction-focused relationships

Content Authority Model:

- "Here's what I know that helps you"

- Original, valuable content
- Proactive expertise sharing
- Relationship-focused approach

The Data Behind the Authority Gap

According to NAR's 2023 Profile of Home Buyers and Sellers, 97% of buyers used the internet in their home search process. Yet when examining where consumers actually find their realtor, the picture becomes stark:

- 25% used a real estate professional they had worked with before for buying, selling, or referrals
- 11% contacted a real estate company directly
- 8% found their agent through a yard sign
- Only 6% found their agent through the internet

This disconnect reveals the authority crisis: while nearly all buyers research online, very few actually discover their agent through digital channels. This suggests that most real estate professionals, despite having some online presence, lack the digital authority necessary to attract new clients through their expertise.

The Invisible Cost of Invisibility

When you don't own your digital authority, you pay in ways that compound over time:

Short-term costs:
- Higher lead acquisition costs
- Price-based competition
- Longer sales cycles
- Constant prospecting pressure

Long-term costs:
- No transferable business equity
- Dependence on others' platforms and algorithms
- Inability to weather market changes
- Limited referral network growth

Your Authority Audit: Where Do You Stand?

Before we explore the solution, take an honest assessment of your current digital authority:

1. *Search yourself + your city.*
   What appears in the first 10 results?
2. *Search for local real estate advice in your market.*
   Do you appear anywhere?
3. *Review your current lead sources.*
   What percentage comes from your expertise vs. paid advertising or referrals?
4. *Examine your client conversations.*
   Do they mention your content, insights, or expertise before meeting you?

If you're not satisfied with these answers, you're ready to build true digital authority.

The Path Forward

The solution isn't abandoning your broker's resources or fighting against industry platforms. It's about building a complementary authority platform that YOU own and control—one that works alongside your existing tools while positioning you as the recognized expert in your market.

In the following chapters, we'll explore exactly how to build this platform and why it's become essential for any real estate professional serious about long-term success and premium positioning.

> *The question isn't whether you need*
> *your own authority platform—*
> *it's whether you'll build it before or after*
> *your competitors do.*

Chapter 2

Website Ownership as Business Asset

The Digital Real Estate You Actually Own

In real estate, you understand the fundamental difference between owning and renting property. When you own, you build equity, control improvements, and benefit from appreciation. When you rent, you're subject to someone else's rules, can lose access at any time, and build no long-term wealth.

The same principle applies to your digital presence, yet most real estate professionals have unknowingly chosen to rent their entire online identity. According to the National Association of Realtors® 2023 Technology Survey, 89% of real estate professionals use social media for real estate purposes, and 77% have some form of website presence. However, the vast majority of this digital activity happens on platforms they don't control—Facebook business pages, LinkedIn profiles, broker-provided websites, and portal listings.

This creates a dangerous dependency that puts your business at the mercy of platform owners, algorithm changes, and policy shifts you cannot predict or control.

The Platform Dependency Crisis

Consider the digital platforms most real estate professionals rely on and the inherent risks each presents:

Social Media Platforms:

Facebook's algorithm changes have dramatically reduced organic reach for business pages. According to various industry studies, organic reach for Facebook business pages has declined from around 16% in 2012 to less than 2% today. This means if you have 1,000 followers, fewer than 20 people typically see your posts without paid promotion.

Instagram's algorithm prioritizes personal connections over business content, making it increasingly difficult for real estate professionals to reach potential clients organically. LinkedIn has similarly adjusted its algorithm to favor individual engagement over promotional business content.

Real Estate Portals:

Real Estate portals control how agent information is displayed and can change their policies at any time. In recent years, these platforms have reduced agent visibility while promoting their own services. Zillow's brief entry into iBuying demonstrated how quickly these companies can shift from lead generation partners to direct competitors.

Broker Websites:

While professionally designed, broker websites present several ownership challenges:

- All SEO equity and domain authority belongs to the brokerage
- Content and lead capture typically benefit the company brand over individual agents
- When you change brokers, you lose all digital equity built on their platform
- Limited customization prevents true personal brand development

The Cost of Platform Dependence

The NAR Member Profile shows that the average real estate professional changes brokerages multiple times during their career. Each transition represents a potential loss of digital equity for agents who've built their online presence primarily on their broker's platform.

Beyond brokerage changes, platform dependence creates several measureable costs:

Acquisition Cost Inflation:

As organic reach declines on social platforms, real estate professionals must increasingly pay for visibility. Facebook advertising costs for real estate have increased significantly over the past five years, with cost-per-click rates rising as competition intensifies and organic reach diminishes.

Revenue Vulnerability:

Real estate professionals dependent on portal leads often find themselves competing primarily on price, as these platforms commoditize agent services. This creates downward pressure on commissions and reduces profit margins.

Limited Differentiation:

Platform-based presence makes it difficult to showcase unique expertise or local knowledge. Portal profiles and social media posts provide minimal space for demonstrating deep market insights or professional specialization.

Building Digital Equity Through Ownership

When you own your website, every piece of content you create builds equity in your personal brand and business. This equity manifests in several measurable ways:

Search Engine Authority:

Consistent, valuable content on your owned domain builds domain authority over time. Unlike social media posts that quickly disappear from feeds, website content remains searchable and continues to attract visitors months or years after publication.

Google's algorithm updates increasingly favor websites that demonstrate expertise and authority in specific topics. Real estate professionals who consistently publish market insights, neighborhood analyses, and educational content on their own domains benefit from improved search rankings and organic traffic growth.

Content Compounding Effect:

Each blog post, market report, or educational resource you publish on your website adds to your total body of work. This creates a compounding effect where your site becomes increasingly valuable as a resource, attracting more visitors and establishing stronger expertise signals.

Email List Ownership:

Unlike social media followers, email subscribers represent a direct communication channel you control. Email marketing consistently delivers higher ROI than social media marketing, with average returns of $42 for every dollar spent according to various industry studies.

Referral Value Enhancement:

When past clients or referral sources want to recommend you, a comprehensive website with valuable content makes their job easier. They can direct prospects to specific resources that demonstrate your expertise, rather than generic social media profiles or broker pages.

The Business Asset Valuation

Professional business appraisers increasingly recognize website traffic, email lists, and content libraries as valuable business assets. While methodologies vary, established websites with consistent traffic and engaged audiences can add significant value to a real estate practice.

Measurable Asset Components:
- Organic search traffic and keyword rankings
- Email subscriber lists with demonstrated engagement
- Content library with proven traffic generation
- Brand recognition and authority in local market
- Lead generation systems and conversion funnels

Transfer and Succession Benefits:
Unlike social media accounts or broker-dependent web presence, owned websites can be transferred, sold, or passed to successors. This creates genuine business equity that extends beyond immediate transaction income.

The AI Era Amplification Effect

The rise of AI in search and content discovery actually strengthens the case for website ownership. AI systems like Google's Search Generative Experience prioritize content from authoritative sources when providing answers to user queries.

Websites with consistent, expert-level content are more likely to be referenced by AI systems as trusted sources. This creates a significant advantage for real estate professionals who have built comprehensive content libraries on their owned platforms versus those relying solely on social media posts or portal profiles.

Additionally, AI tools enable more sophisticated content

creation and optimization strategies for website owners, while platform-dependent real estate professionals remain subject to algorithm changes they cannot influence or predict.

Risk Mitigation Through Diversification

Owning your website doesn't mean abandoning other platforms—it means creating a diversified digital strategy with your website as the foundation. This approach provides several risk mitigation benefits:

Algorithm Independence:
While social media algorithms can dramatically reduce your reach overnight, owned website content remains accessible to anyone who searches for it or visits directly.

Platform Policy Protection:
Social media platforms regularly change their terms of service, advertising policies, and content guidelines. Website owners maintain control over their content and communication methods.

Competitive Insulation:
As real estate portals and platforms increasingly compete with agents, having an independent web presence provides protection from potential policy changes that could limit agent visibility or lead access.

Implementation Realities
The transition from platform dependence to website ownership doesn't happen overnight, but it doesn't require abandoning existing marketing channels. The most effective approach involves gradually building your owned platform while maintaining strategic use of other channels to drive traffic and engagement.

Immediate Benefits:
Even a basic website with regular content updates begins building SEO equity and providing a professional destination for referrals and leads.

Long-term Compound Growth:
Website authority and traffic typically grow exponentially over time, with each piece of quality content contributing to overall site value and search visibility.

Competitive Differentiation:
In markets where most real estate professionals rely on identical platform-based marketing, owned content platforms provide significant differentiation and professional credibility.

The Ownership Mindset Shift

Moving from platform dependence to website ownership requires a fundamental mindset shift from renting digital real estate to owning it. This shift affects not just your marketing tactics, but your entire approach to professional development and business building.

Instead of asking "How can I get more followers?" the question becomes "How can I build a valuable resource that attracts my ideal clients?" Instead of competing for algorithm favor, you focus on creating genuine value that establishes lasting authority in your market. This ownership mindset aligns perfectly with the expertise and local knowledge that successful real estate professionals already possess—it simply provides a platform to showcase and leverage that expertise effectively.

Chapter 3

Beyond IDX - The AI-Era Content Authority Model

The Listings Limitation

Most real estate professionals websites focus heavily on property listings through IDX integration, operating under the assumption that showing available inventory drives business. While property search functionality serves an important purpose, it creates a fundamental positioning problem: it makes you a facilitator of listings rather than a source of expertise.

According to NAR's Profile of Home Buyers and Sellers, 97% of buyers used the internet during their home search, but only 17% found the home they purchased online. The disconnect reveals a critical insight: buyers use the internet primarily for research and education, not just property browsing.

This research phase represents the largest missed opportunity for most real estate professionals. While buyers are actively seeking market insights, neighborhood information, and process guidance, most agent websites offer little beyond property listings that duplicate what's available on major portals—often with inferior search functionality.

The Authority Gap in Real Estate Marketing

Traditional real estate marketing focuses on transaction facilitation: "*I can help you buy or sell.*" In contrast, authority-based marketing focuses on knowledge demonstration: "*Here's what I know that helps you make better decisions.*"

The NAR Technology Survey shows that while 77% of real estate professionals maintain some form of website, very few use their sites to consistently demonstrate market expertise. Most websites function as digital business cards rather than authority platforms.

This creates significant opportunity for real estate professionals willing to shift their approach from transaction marketing to expertise marketing. In markets where dozens of agents offer similar services, the agent who consistently demonstrates superior knowledge and insights gains a substantial competitive advantage.

Google's E-E-A-T Framework and Real Estate

Google's emphasis on Experience, Expertise, Authoritativeness, and Trustworthiness (E-E-A-T) in search rankings has particular relevance for real estate professionals. The framework prioritizes content from sources that demonstrate:

> **Experience**: First-hand knowledge of topics being discussed
>
> **Expertise**: Deep understanding and specialized knowledge
>
> **Authoritativeness**: Recognition as a credible source in the field
>
> **Trustworthiness**: Reliability and transparency of information

Real estate transactions represent one of the largest financial decisions most people make, placing them squarely in Google's "Your Money or Your Life" (YMYL) category. This means search algorithms apply particularly strict standards when evaluating real estate content quality and source credibility.

Real estate professionals who consistently publish content demonstrating local market experience, transaction expertise, and industry knowledge benefit from improved search visibility. Conversely, generic or promotional content receives lower rankings, regardless of SEO optimization.

The Multi-Platform Authority Model

The modern digital landscape requires moving beyond single-platform thinking to create integrated authority across multiple channels. Your website serves as the comprehensive authority hub, while other platforms function as distribution channels that amplify your expertise and drive traffic back to your owned content.

Content Hub Strategy (Your Website):

Your website hosts comprehensive, in-depth content that showcases your complete expertise. This includes detailed market analyses, neighborhood guides, educational resources, and process explanations that demonstrate deep knowledge and local experience.

Distribution Channel Strategy (Other Platforms):

Social media, video platforms, and email marketing serve as channels to share insights, engage in discussions, and direct interested prospects to your comprehensive website content. This approach leverages each platform's strengths while building long-term equity in your owned platform. Social media provides immediacy and engagement, video platforms offer visual storytelling, and email enables direct relationship building—all while driving authority-seeking prospects to your website for detailed expertise.

Content Categories That Build Authority

Effective authority-building content for real estate professionals falls into four primary categories, each serving different audience needs and search intents:

Market Intelligence and Data Interpretation

Raw market data is widely available, but interpretation requires expertise. Content that analyzes trends, explains implications, and provides context demonstrates professional knowledge beyond basic information sharing. Examples include monthly market reports that explain what statistics mean for buyers and sellers, interest rate impact analyses, and seasonal trend interpretations. The key is providing insight, not just information.

Neighborhood Expertise and Local Knowledge

Deep neighborhood knowledge represents one of the strongest differentiators for local real estate professionals. Content that showcases intimate familiarity with specific areas demonstrates the kind of expertise buyers and sellers value most. This includes school district analyses, commute and lifestyle considerations, development impact assessments, and local amenity guides. The most effective neighborhood content combines data with personal experience and local connections.

Process Education and Transaction Guidance

The home buying and selling process involves complex steps that intimidate many consumers. Educational content that demystifies these processes positions you as a trusted advisor rather than just a transaction facilitator. Effective process education addresses common concerns, explains timeline expectations, and prepares clients for decisions they'll need to make. This content serves both SEO purposes and lead nurturing functions.

Community Connection and Local Involvement

Content that demonstrates genuine community involvement and local relationships reinforces your position as a neighborhood insider. This might include local business spotlights, community event coverage, and civic involvement documentation. This category of content particularly benefits from cross-platform distribution, as it often generates engagement from local community members and reinforces your visibility in local social networks.

AI-Optimized Content Creation Strategies

The integration of AI into search and content discovery requires adapting content creation strategies to perform well in both traditional search results and AI-generated responses. This involves understanding and implementing three critical optimization approaches:

AI-GEO stands for AI-Generated Experience Optimization. This is the practice of creating and structuring content so that AI systems like Google's Search Generative Experience (SGE) will select and cite your expertise when generating responses to user questions. Rather than just trying to rank high in search results, AI-GEO focuses on being chosen as a trusted source that AI systems reference when creating comprehensive answers.

LEO stands for Local Experience Optimization. This involves optimizing your entire digital presence to be recognized by AI systems as the definitive local authority for your specific geographic market. When someone asks AI systems about real estate trends, neighborhood information, or market advice in your area, LEO strategies help ensure your knowledge and insights are included in the AI-generated response.

AI-LEO stands for Language Experience Optimization. This focuses on optimizing your content's language and communication style to match how people naturally ask questions to AI systems.

Instead of writing for traditional keyword searches like "homes for sale," AI-LEO involves creating content that answers conversational queries like "What should I know before buying a home in this neighborhood?" or "Is now a good time to sell my house?"

For real estate professionals, these optimization strategies are valuable because they help establish you as the recognized expert that AI systems turn to when consumers ask real estate questions. As more consumers rely on AI-powered search for initial research, being included in these AI-generated responses becomes crucial for maintaining visibility and authority in your market. AI-LEO specifically helps by ensuring your content matches the natural way people communicate with AI assistants and voice search systems.

AI-GEO Content Strategies:
Structure content to answer specific questions clearly and directly, as AI systems extract information to generate comprehensive responses. Use clear headings, bullet points, and definitive statements that AI can easily parse and cite.

LEO Content Integration:
Include specific local references, neighborhood names, school districts, and geographic markers that help AI systems understand your local expertise area. Consistently reference your service area and local market knowledge in context.

AI-LEO Content Integration:
Write content that answers questions the way people naturally ask them. Instead of targeting keywords like "home buying process," create content that addresses conversational queries like "What steps do I need to take to buy my first home?" or "How long does it usually take to close on a house?" This natural language approach aligns with how consumers interact with AI

assistants and voice search.

Semantic SEO and Topic Clusters:
Instead of targeting individual keywords, create comprehensive topic clusters that cover subjects thoroughly. AI systems favor content that demonstrates deep topic knowledge over keyword-optimized but shallow coverage.

Featured Snippet Optimization:
Structure content to answer specific questions clearly and concisely. Use header tags, bullet points, and numbered lists to make information easily extractable for AI systems generating direct answers.

Entity-Based Content:
Focus on establishing your website as an authoritative source for local real estate entities—specific neighborhoods, school districts, developments, and market segments. Consistent, detailed coverage of these entities builds both traditional SEO authority and AI-GEO relevance.

Cross-Reference Integration:
Link related content pieces to create comprehensive resource networks. AI systems favor websites that provide thorough, interconnected information over isolated articles, and this linking helps establish topical authority for LEO purposes.

The Consistency Imperative

Authority building requires consistent content creation over extended periods. Sporadic publishing patterns undermine credibility and provide insufficient material for search engines to establish topical expertise. Industry research suggests that websites publishing weekly content see significantly better search performance than those publishing monthly or less frequently. However, quality remains paramount—consistent publication of valuable content outperforms frequent publication of low-value material.

Content Calendar Development:
Successful content authority requires systematic planning and execution. This includes identifying recurring content opportunities (monthly market reports, quarterly trend analyses), seasonal topics (spring buying season, year-end tax considerations), and evergreen educational content.

Repurposing for Efficiency:
Comprehensive website content can be repurposed across multiple platforms and formats. A detailed neighborhood analysis might become a blog post, social media series, email newsletter content, and video presentation topics.

Measuring Authority Building Success
Traditional SEO metrics like keyword rankings remain relevant, but authority building requires broader measurement approaches that capture expertise demonstration and audience engagement.

Search Performance Indicators:
- Organic traffic growth to educational and informational content
- Featured snippet appearances for local real estate topics
- Brand mention tracking in local online discussions
- Position improvements for expertise-based search terms

Engagement and Authority Signals:
- Time spent on educational content pages

- Email subscription rates from content consumption
- Social sharing and discussion generation
- Direct traffic increases indicating brand recognition

Business Impact Metrics:
- Lead quality improvements from content-driven prospects
- Consultation request increases attributable to content consumption
- Referral source mentions of content or expertise
- Premium positioning success in competitive situations

Integration with Traditional Marketing

The content authority model doesn't replace traditional real estate marketing—it enhances and amplifies existing efforts. Client testimonials gain more credibility when supported by demonstrated expertise. Networking becomes more effective when you can direct contacts to valuable resources. Referral partners have specific content to share when recommending your services.

Referral Enhancement:
Past clients and referral sources can easily share specific content pieces that demonstrate your expertise, making their recommendations more compelling and specific.

Networking Amplification:
Professional relationships benefit when you can provide valuable insights and resources rather than just requesting referrals or business.

Client Education:
Comprehensive content resources help educate clients through-

out the transaction process, improving satisfaction and reducing repetitive explanations.

The Long-Term Competitive Advantage

Content authority building creates compound advantages that strengthen over time. Each piece of quality content adds to your total expertise demonstration, improving search visibility and professional positioning. This creates increasingly difficult competitive barriers as your content library and authority signals grow. Unlike advertising or promotional marketing that requires continuous investment for ongoing results, content authority builds lasting value. Educational content remains relevant and continues attracting prospects long after publication, creating an expanding asset base rather than recurring expenses.

The real estate professionals who begin building content authority now will have significant competitive advantages as more professionals recognize the importance of expertise demonstration in the AI-influenced search landscape.

Chapter 4:

The Psychology of Trust in Real Estate

The Trust Decision Framework

Real estate transactions involve complex decisions with significant financial and emotional stakes. According to NAR data, the median existing home sale price reached $406,700 in 2023, representing the largest single purchase most consumers will make. Beyond the financial magnitude, homes carry deep emotional significance as the foundation for family life, security, and personal identity.

This combination of high stakes and emotional investment creates a unique decision-making environment where trust becomes the primary factor in agent selection. Understanding how trust develops in the digital age is essential for real estate professionals seeking to attract and retain quality clients.

The Traditional Trust Model vs. Digital Reality

Traditional real estate trust building relied heavily on personal referrals, community presence, and face-to-face interactions. While these elements remain important, consumer research patterns have fundamentally changed how initial trust impressions form.

The NAR Profile of Home Buyers and Sellers shows that 97% of buyers used the internet during their home search process. However, this research isn't limited to property browsing—consumers extensively research potential agents online before making contact. This means initial trust impressions increasingly form through digital interactions rather than personal meetings.

The Pre-Contact Research Phase:
Modern consumers typically research potential agents for weeks or months before making contact. During this phase, they're evaluating competence, local knowledge, communication style, and professional credibility through available online content and presence.

This research-intensive approach means that by the time a consumer contacts an agent, they've already formed preliminary trust judgments based on digital evidence of expertise and professionalism.

The Psychology of Expertise-Based Trust

Trust in professional relationships develops through several psychological mechanisms, with demonstrated competence playing a particularly crucial role in high-stakes decisions like real estate transactions.

Competence-Based Trust:
Consumers assess professional competence through evidence of knowledge, experience, and successful outcomes. In the digital environment, this evidence must be readily discoverable and consistently demonstrated through content and online presence.

Consistency and Reliability:
Trust develops through consistent behavior patterns over time. For real estate professionals, this means regular demonstration of market knowledge, reliable communication, and predictable professional standards across all interactions.

Transparency and Authenticity:
Modern consumers expect transparency in professional relationships. Real estate professionals who openly share market insights, explain process details, and acknowledge market challenges build stronger trust than those who maintain purely promotional messaging.

Social Proof and Validation:
While testimonials provide important social proof, the most compelling trust signals combine client feedback with independent evidence of expertise and professional recognition.

The Content Trust-Building Mechanism
Regular publication of valuable, accurate content creates trust through multiple psychological pathways:

Demonstrated Expertise:
Consistent sharing of market insights, process explanations, and local knowledge provides ongoing evidence of professional competence. This cumulative demonstration often carries more weight than claims of expertise or credentials alone.

Accessibility and Helpfulness:
Content that answers common questions and addresses consumer concerns demonstrates a service orientation that builds confidence in the professional relationship.

Transparency in Communication:
Educational content that explains market realities, process complexities, and potential challenges shows honesty and builds realistic expectations—key components of trustworthy relationships.

Professional Investment:

The time and effort required to create valuable content signals professional commitment and seriousness, differentiating serious professionals from those seeking quick transactions.

Why Traditional "About Me" Approaches Fail

Most real estate professional's websites feature traditional "About Me" sections that focus on credentials, achievements, and personal information. While this information has value, it fails to address the primary trust question in consumers' minds: "Can this person help me succeed in my real estate goals?"

The Self-Focus Problem:

Traditional about pages center on the agent's accomplishments rather than the client's needs and concerns. This approach misses opportunities to demonstrate understanding of client challenges and market complexities.

Credential Overemphasis:

While certifications and awards have value, they don't address consumers' primary concerns about local knowledge, market timing, negotiation ability, and process management.

Generic Messaging:

Standard phrases like "committed to excellence" and "working hard for you" appear on countless agent websites, creating no differentiation or specific trust signals.

Trust-Building Content Strategies

Effective trust building through content requires shifting focus from self-promotion to value demonstration and client education.

Problem-Solving Orientation:
Content that addresses specific challenges buyers and sellers face demonstrates understanding of client needs and professional problem-solving ability.

Local Market Insight Sharing:
Regular analysis of local market conditions, neighborhood developments, and economic factors shows deep market knowledge and ongoing professional engagement.

Process Education and Preparation:
Content that prepares clients for transaction steps, decision points, and potential challenges builds confidence in professional guidance and reduces anxiety about complex processes.

Honest Market Assessment:
Balanced analysis that acknowledges both opportunities and challenges in current market conditions demonstrates integrity and builds realistic expectations.

The Vulnerability and Authenticity Factor

Research in trust psychology shows that appropriate vulnerability and authenticity strengthen professional relationships, particularly in high-stakes situations.

Professional Vulnerability:
Sharing challenges overcome, lessons learned, and market complexities navigated demonstrates human authenticity while maintaining professional competence signals.

Market Reality Acknowledgment:
Honest discussion of market challenges, timing uncertainties, and decision complexities builds trust through transparency

rather than oversimplified optimism.

Client Success Focus:
Sharing client success stories that focus on problem-solving and positive outcomes rather than personal achievements demonstrates client-centered orientation.

Digital Trust Signals and Authority Indicators
In the digital environment, specific elements serve as trust signals that influence consumer confidence and agent selection:

Content Depth and Quality:
Comprehensive, well-researched content indicates professional knowledge and commitment to providing value rather than just generating leads.

Regular Publication Patterns:
Consistent content creation demonstrates ongoing market engagement and professional activity rather than sporadic marketing efforts.

Local Knowledge Demonstration:
Specific, detailed content about neighborhoods, schools, developments, and local market factors provides evidence of genuine local expertise.

Professional Presentation:
Well-designed, professional websites and thoughtfully written content signal attention to detail and professional standards.

Engagement and Responsiveness:
Active participation in online discussions, prompt responses to questions, and engagement with community topics demonstrate

accessibility and client service orientation.

The Compound Effect of Trust Building

Trust building through content creates compound advantages that strengthen over time:

Accumulated Evidence:

Each piece of valuable content adds to the total evidence of expertise and professionalism, creating stronger overall trust impressions.

Referral Amplification:

Clients who have experienced demonstrated expertise through content are more likely to provide specific, enthusiastic referrals that mention professional knowledge and helpfulness.

Quality Client Attraction:

Strong digital trust signals tend to attract more serious, qualified prospects who appreciate professional expertise rather than just seeking lowest-cost services.

Reduced Sales Cycle:

Prospects who develop trust through content consumption often require less convincing during initial consultations, leading to faster agreement and better working relationships.

Trust Maintenance and Relationship Development

Trust building doesn't end with client acquisition—it continues throughout the relationship and beyond:

Ongoing Education:

Continued sharing of relevant market information and process updates maintains trust and demonstrates ongoing professional development.

Transparent Communication:
Regular updates about market conditions, process status, and potential challenges maintain trust through transparency and proactive communication.

Post-Transaction Value:
Continued provision of valuable content and market insights after transaction completion builds long-term relationships and referral opportunities.

Measuring Trust Development
While trust is primarily qualitative, several indicators can help measure trust-building effectiveness:

Engagement Quality:
Length of website visits, content consumption patterns, and email engagement rates indicate developing interest and confidence.

Inquiry Quality:
Questions that reference specific content or demonstrate familiarity with your expertise indicate successful trust building through content consumption.

Referral Feedback:
Comments from referral sources about the specific aspects of your service they recommend provide insight into trust factors that resonate most strongly.

Client Decision Speed:
Prospects who have consumed significant content often make

faster decisions to work together, indicating pre-developed trust through expertise demonstration.

The Professional Relationship Advantage

Trust built through demonstrated expertise creates fundamentally different client relationships than those based primarily on personal chemistry or price competition:

Advisory Positioning:
Clients who recognize expertise are more likely to seek advice and guidance rather than just transaction services.

Premium Value Recognition:
Strong trust in professional competence supports premium pricing as clients understand the value of expertise in high-stakes transactions.

Reduced Stress Relationships:
Clients confident in professional expertise experience less anxiety during complex transactions, leading to more positive working relationships.

Long-Term Loyalty:
Trust based on demonstrated competence creates stronger client loyalty than relationships based primarily on personal rapport or convenience.

The psychology of trust in real estate requires understanding both the emotional stakes involved and the digital research patterns that shape initial impressions. Real estate professionalss who systematically build trust through content and expertise demonstration create sustainable competitive advantages that strengthen over time.

Chapter 5

Lead Capture That Actually Works

The Value-First Lead Generation Revolution

Traditional real estate lead capture relies on gate-keeping property information or using generic contact forms with messages like "Contact me for more details" or "Schedule your free consultation." This approach worked when consumers had limited access to real estate information, but it fails in today's information-rich environment where buyers and sellers can access property data, market statistics, and educational content from multiple sources.

According to NAR's Technology Survey, real estate consumers are more informed than ever before, with 97% using the internet during their home search process. These sophisticated consumers expect value before they're willing to share contact information. They want to understand what specific expertise you bring to their situation before committing to a business relationship.

The most successful modern lead capture strategies operate on a value-first principle: provide genuine utility in exchange for contact information, then continue delivering value to build trust and relationship over time.

Understanding the Modern Consumer's Information Journey

Today's real estate consumers follow a predictable information-gathering pattern before engaging with agents:

Research Phase (Weeks to Months):

Consumers research neighborhoods, market conditions, process requirements, and timing considerations long before they're ready to transact. During this phase, they're seeking educational content and expert insights, not sales presentations.

Evaluation Phase (Days to Weeks):
Once consumers decide to move forward with buying or selling, they research potential agents by examining online presence, expertise demonstrations, and professional credibility indicators.

Engagement Phase (Hours to Days):
Only after thorough research do consumers contact agents, and by this point they've often formed preliminary preferences based on perceived expertise and professionalism.

Understanding this journey reveals why value-based lead capture works: it aligns with consumers' natural research patterns and provides utility during the phase when they're most receptive to expert guidance.

The Psychology of Value Exchange

Effective lead capture succeeds because it creates a fair value exchange that feels beneficial to both parties. Consumers receive something immediately useful, while real estate professionals gain contact information and an opportunity to demonstrate expertise.

Reciprocity Principle:
When you provide genuine value without immediate expectation of return, it triggers psychological reciprocity that makes consumers more receptive to future communication and more likely to consider your services when ready to transact.

Authority Demonstration:
Well-crafted lead magnets showcase your knowledge and local expertise, providing evidence of competence that builds trust before direct interaction.

Relationship Initiation:
Value-based lead capture begins the relationship on a positive foundation where you're seen as helpful and knowledgeable rather than pushy or sales-focused.

High-Value Lead Magnets for Real Estate

Effective lead magnets for real estate professionals provide specific, actionable information that addresses genuine consumer needs and demonstrates local market expertise:

Neighborhood Buyer's Guides
Comprehensive guides that go beyond basic statistics to provide insider insights about specific neighborhoods. These might include:

- Commute patterns and transportation options for different lifestyle needs
- Local amenity assessments including shopping, dining, and recreation
- Development pipeline information and potential neighborhood changes
- Seasonal market patterns and optimal timing considerations
- Price trend analysis with context about driving factors

The key is providing information that demonstrates intimate neighborhood knowledge rather than data available through public sources.

Market Timing and Strategy Guides

Educational resources that help consumers understand current market conditions and make strategic decisions:

- "*Is Now the Right Time to Sell?*" assessment tools with local market factors
- "*First-Time Buyer's Market Navigation Guide*" for current conditions
- "*Investment Property Analysis Worksheet*" with local rental market data
- "*Downsizing Strategy Guide*" addressing local market considerations
- "*New Construction vs. Existing Home*" comparison with local examples

These guides position you as a strategic advisor rather than just a transaction facilitator.

Process Education and Preparation Resources

Detailed explanations of complex processes that reduce consumer anxiety and demonstrate professional competence:

- "*Complete Home Buying Timeline*" with local market specifics
- "*Selling Preparation Checklist*" with market-specific recommendations
- "*Understanding Real Estate Contracts*" guide with state and local considerations
- "*Closing Process Walkthrough*" addressing common concerns and timeline
- "*Negotiation Strategies*" guide for current market conditions

Process education builds confidence in your ability to guide clients successfully through complex transactions.

Financial Planning and Analysis Tools

Interactive tools or comprehensive guides that help consumers make informed financial decisions:

- *"True Cost of Homeownership Calculator"* with local tax and insurance data
- *"Rent vs. Buy Analysis"* customized for local market conditions
- *"Home Equity Strategy Guide"* for existing homeowners
- *"Moving Cost Calculator"* with local service provider information
- *"Property Tax Impact Analysis"* for different neighborhoods and home values

Financial guidance demonstrates sophistication and advisory capability that sets you apart from transaction-focused competitors.

Lead Magnet Creation Best Practices

Successful lead magnets share several characteristics that maximize both download rates and lead quality:

Specific Problem Solving:

Address particular challenges or questions rather than providing generic information. *"Understanding [Your City]'s Zoning Changes"* performs better than *"Guide to Local Zoning"*

Local Market Integration:

Include specific local information, data, and examples that couldn't be found in generic real estate guides. This demonstrates genuine local expertise and provides unique value.

Professional Presentation:

High-quality design and professional writing signal attention to detail and competence. Poor presentation undermines credibility regardless of content quality.

Actionable Information:
Provide concrete steps, checklists, or decision frameworks rather than just general advice. Consumers should be able to immediately apply what they've learned.

Update Capability:
Create lead magnets that can be regularly updated with current market information, maintaining relevance and providing opportunities for re-engagement.

Landing Page Optimization for Trust and Conversion

The landing page that promotes your lead magnet is often a prospect's first direct interaction with your brand, making optimization crucial for both conversion and trust building:

Clear Value Proposition:
Immediately explain what specific problem the resource solves and why your local expertise makes it valuable. Avoid generic benefits like "insider knowledge" in favor of specific outcomes.

Trust Signal Integration:
Include professional credentials, local market experience, and social proof that reinforces your credibility and expertise.

Minimal Friction Design:
Request only essential information (typically name and email) to reduce abandonment. Additional qualification can occur through follow-up communication.

Mobile Optimization:

Ensure landing pages function perfectly on mobile devices, as significant traffic comes from consumers researching on phones and tablets.

Thank You Page Strategy:

Use confirmation pages to begin the relationship immediately with additional value, social media connections, or consultation scheduling options.

Email Nurture Sequence Development

The period immediately following lead magnet download is crucial for relationship development and trust building:

Immediate Delivery and Welcome:

Deliver the promised resource immediately and welcome subscribers with a personal message that sets expectations for future communication.

Educational Series:

Follow up with a series of educational emails that continue demonstrating expertise while avoiding sales pressure. Share market insights, process tips, and local knowledge.

Social Proof Integration:

Include client success stories, testimonials, and professional recognition in nurture emails to build credibility over time.

Value-Added Resources:

Occasionally provide additional resources, tools, or insights that weren't part of the original lead magnet, reinforcing the value of staying subscribed.

Soft Call-to-Action Integration:
Include gentle invitations for consultation or conversation without high-pressure sales language. Focus on helping rather than selling.

Segmentation and Personalization Strategies
Different types of prospects have different information needs and timeline considerations:

Buyer vs. Seller Segmentation:
Create separate nurture sequences for buyers and sellers, as their information needs and timeline patterns differ significantly.

Timeframe Consideration:
Segment leads based on indicated timeline (immediate need vs. future planning) to provide appropriately timed information and avoid overwhelming future-focused prospects.

Geographic Focus:
For real estate professionals serving multiple areas, segment by neighborhood or region of interest to provide hyper-local information and maintain relevance.

Experience Level:
First-time buyers need different information than experienced investors or repeat buyers. Tailor content complexity and focus areas accordingly.

Integration with Content Marketing Strategy
Lead magnets work most effectively when integrated with ongoing content marketing efforts:

Blog Content Connection:

Create blog posts that address related topics and naturally lead to lead magnet downloads for deeper information.

Social Media Promotion:

Share insights from lead magnets on social platforms to demonstrate value and drive traffic to landing pages.

Video Content Integration:

Create video content that summarizes key points from written resources, appealing to different learning preferences while driving downloads.

Email Newsletter Integration:

Regular newsletter content can reference and promote relevant lead magnets to existing subscribers who might benefit from specific resources.

Measuring Lead Magnet Effectiveness

Success measurement should focus on both quantity and quality metrics:

Conversion Rate Optimization:

Track landing page conversion rates and test different headlines, value propositions, and designs to improve performance.

Lead Quality Assessment:

Monitor which lead magnets generate prospects who eventually become clients, focusing efforts on the most effective resources.

Engagement Tracking:

Measure email open rates, click-through rates, and unsubscribe patterns to understand which content resonates most strongly.

Attribution Analysis:
Track which lead magnets contribute to consultation bookings and eventual client relationships to calculate true ROI.

Advanced Lead Capture Strategies
As your content authority grows, consider more sophisticated lead capture approaches:

Interactive Tools and Calculators:
Develop online tools that provide personalized analysis based on user input, creating higher engagement and perceived value.

Webinar and Workshop Integration:
Use educational presentations as lead capture mechanisms while demonstrating expertise through live interaction.

Multi-Step Lead Magnets:
Create resource sequences that build engagement over time, such as email courses or weekly market updates.

Community Access: Offer access to private Facebook groups or online communities focused on local real estate discussion and education.

The Long-Term Relationship Building Effect
Value-first lead capture creates fundamentally different prospect relationships than traditional methods:

Educational Positioning:
Prospects see you as an educator and advisor rather than a salesperson, creating more receptive audience for future communication.

Trust Development:

Consistent value delivery builds trust over time, making prospects more likely to choose your services when ready to transact.

Referral Generation:

Satisfied lead magnet users often refer others even before becoming clients themselves, expanding your reach through word-of-mouth marketing.

Market Positioning:

Comprehensive lead magnet libraries position you as the definitive local expert, creating competitive differentiation that's difficult to replicate.

Effective lead capture in today's market requires shifting from gate-keeping information to providing genuine value that demonstrates expertise and builds trust. This approach attracts higher-quality prospects while establishing the foundation for long-term client relationships based on perceived competence rather than just personal rapport.

Chapter 6

The AI Era Multi-Platform Strategy

Beyond Single-Platform Thinking

The most successful real estate professionals in today's digital landscape don't rely on any single platform for their marketing efforts. Instead, they create integrated systems where each platform serves a specific purpose while contributing to overall authority building and business growth.

This approach has become even more critical as AI systems increasingly evaluate authority across multiple digital touchpoints. Google's algorithm updates, social media algorithm changes, and the rise of AI-powered search results all point toward the same conclusion: diverse, consistent expertise demonstration across platforms creates stronger authority signals than single-platform excellence.

According to various industry studies, consumers interact with brands across an average of 6-8 touchpoints before making purchasing decisions. For high-stakes real estate transactions, this number is often higher, making multi-platform consistency essential for building the trust necessary to secure client relationships.

The Hub and Spoke Authority Model

The most effective multi-platform strategy operates on a hub and spoke model where your website serves as the comprehensive authority hub while other platforms function as specialized distribution channels:

Website Hub Functions:

- Comprehensive content library demonstrating complete expertise
- Lead capture systems with detailed resource offerings
- Professional presentation and trust signal optimization
- SEO optimization for long-term organic traffic growth
- Email list building and nurture system management

Platform Spoke Functions:

- Social media for community engagement and content distribution
- Video platforms for visual expertise demonstration
- Email marketing for direct relationship building
- Professional networks for industry connection and credibility
- Local platforms for community involvement and visibility

This model ensures that all marketing efforts ultimately build equity in your owned platform while leveraging each channel's unique strengths and audience characteristics.

Platform-Specific Strategy Development

Each platform requires tailored content and engagement strategies that align with user expectations while supporting overall authority building goals:

LINKEDIN: PROFESSIONAL AUTHORITY BUILDING

LinkedIn serves as the primary platform for demonstrating professional expertise and industry knowledge to both potential clients and referral sources.

Content Strategy:

Share market analyses, industry insights, and professional observations that showcase deep knowledge and thoughtful perspective. Focus on content that other professionals would find valuable and shareworthy.

Engagement Approach:

Participate in real estate and local business discussions, providing expert insights without overt self-promotion. Comment thoughtfully on industry news and market developments.

Network Development:

Connect with other local professionals, past clients, and industry contacts to build a referral network while maintaining visibility among potential clients who research agents on LinkedIn.

FACEBOOK: COMMUNITY CONNECTION AND LOCAL ENGAGEMENT

Facebook's strength lies in local community building and demonstrating neighborhood involvement and knowledge.

Content Strategy:

Share local market updates, community event information, and neighborhood insights that demonstrate intimate local knowledge and community involvement.

Group Participation:

Actively participate in local Facebook groups, homeowner associations, and neighborhood pages by providing helpful information without aggressive self-promotion.

Video Integration:

Use Facebook's video features for market updates, neighborhood spotlights, and educational content that can be easily shared within local communities.

INSTAGRAM: VISUAL STORYTELLING AND BEHIND-THE-SCENES

Instagram excels at humanizing your professional brand and showcasing the visual aspects of real estate expertise.

Content Strategy:

Share high-quality property photos, neighborhood highlights, and behind-the-scenes glimpses of your professional process. Use Stories for timely market updates and quick tips.

Local Hashtag Strategy:

Utilize location-based hashtags and local community tags to increase visibility among users researching specific neighborhoods or considering relocation.

Professional Lifestyle:

Balance professional content with appropriate personal insights that build relatability while maintaining professional credibility.

YOUTUBE: EDUCATIONAL AUTHORITY AND SEO BENEFITS

YouTube serves dual purposes as both a content platform and search engine, offering significant SEO benefits for authority building.

Content Strategy:

Create educational videos addressing common buyer and seller questions, market analysis presentations, and neighborhood tour content that demonstrates local expertise.

SEO Integration:

Optimize video titles, descriptions, and tags for local real estate searches. Include links back to relevant website content for traffic generation.

Consistency and Series:
Develop regular video series like monthly market updates or neighborhood spotlights to build subscriber loyalty and consistent content output.

TikTok: Trend Integration and Younger Demographics
While not primary for most real estate professionals, TikTok can effectively reach younger demographics and demonstrate personality and expertise in engaging formats.

Content Strategy:
Share quick tips, market insights, and first-time buyer education in TikTok's short, engaging format. Participate in relevant trends while maintaining professional credibility.

Educational Focus:
Use the platform's educational content trends to share real estate knowledge in accessible, entertaining formats that appeal to younger potential clients.

Cross-Platform Content Distribution Strategy
Effective multi-platform marketing requires systematic content creation and distribution that maximizes efficiency while maintaining platform-specific optimization:

Content Creation Hierarchy:
Start with comprehensive content on your website, then create platform-specific adaptations that drive traffic back to the full resource.

Repurposing Framework:
- Blog post → LinkedIn article → Facebook post → Instagram carousel → YouTube video → TikTok highlights

- Market report → Email newsletter → Social media series → Video presentation →Podcast discussion topics

Platform-Specific Optimization:
Adapt content format, length, and style to match each platform's user expectations while maintaining consistent messaging and expertise demonstration.

Traffic Direction:
Include clear calls-to-action that direct engaged users to your website for more comprehensive information and lead capture opportunities.

AI-Era Authority Signal Amplification
Modern search algorithms and AI systems evaluate authority through multiple signals that extend beyond traditional SEO metrics. Understanding AI-GEO and LEO becomes crucial in this multi-platform context:

Cross-Platform AI-GEO Consistency:
AI systems that generate responses to user queries (like Google's SGE) evaluate expertise across multiple platforms. When your local market insights appear consistently across your website, LinkedIn articles, and video content, AI systems gain confidence in your expertise and are more likely to reference your knowledge in generated responses.

Multi-Platform LEO Signals:
Local Experience Optimization benefits significantly from consistent local references across platforms. When you discuss the same neighborhoods, school districts, and local market conditions on your website, social media, and video content, AI systems better understand your geographic expertise area.

Engagement Quality:
Meaningful discussions and interactions across platforms indicate genuine expertise and community value, factors that AI systems increasingly consider when determining authority for both general topics (AI-GEO) and local expertise (LEO).

Content Citation and Reference:
When your content gets shared, referenced, or discussed across platforms, it creates authority signals that search algorithms and AI systems recognize and reward. This is particularly important for LEO, as local discussions and references strengthen geographic authority signals.

Brand Mention Tracking:
AI systems monitor brand mentions and discussions across platforms, making reputation management and positive engagement crucial for authority building. Local mentions and community discussions are especially valuable for LEO optimization.

Integration Workflow Development

Successful multi-platform marketing requires systematic workflows that ensure consistent execution without overwhelming time demands:

Content Calendar Planning:
Develop monthly content calendars that identify key topics, platform-specific adaptations, and distribution schedules across all channels.

Batch Content Creation:
Create multiple pieces of related content simultaneously, then distribute across platforms over time to maintain consistency without constant content creation pressure.

Engagement Schedule:
Establish regular times for platform engagement, responding to comments, and participating in relevant discussions to maintain active presence.

Performance Monitoring:
Track engagement, traffic generation, and lead production across platforms to optimize time investment and content strategy.

Email Marketing as the Unifying Channel
Email marketing serves as the crucial connection between all other marketing efforts, providing direct communication with your most engaged audience:

Cross-Platform Integration:
Use email to promote content published on other platforms while driving subscribers back to your website for comprehensive resources.

Audience Segmentation:
Segment email lists based on content interests, platform preferences, and engagement patterns to provide personalized communication.

Value Delivery:
Provide email-exclusive insights and early access to content to reward subscriber loyalty and maintain high engagement rates.

Relationship Development:
Use email's direct nature to build personal connections and provide consultation opportunities for engaged prospects.

Community Engagement and Local Authority

Multi-platform strategy must include significant focus on local community engagement that reinforces your position as a neighborhood expert:

Local Event Participation:

Document and share community involvement across platforms to demonstrate genuine local connection and investment.

Local Business Relationships:

Collaborate with other local businesses for cross-promotion and community content that showcases neighborhood knowledge.

Civic Involvement:

Participate in local government meetings, community planning discussions, and neighborhood association activities to build credibility and content opportunities.

Local Media Engagement:

Seek opportunities for local media interviews, guest articles, and expert commentary that can be shared across all platforms.

Measuring Multi-Platform Success

Effective measurement requires tracking both individual platform performance and overall integrated campaign results:

Platform-Specific Metrics:

Monitor engagement rates, reach, and conversion metrics for each platform to optimize content and posting strategies.

Cross-Platform Attribution:

Track how prospects interact with content across multiple platforms before converting to leads or clients.

Authority Building Indicators:
Monitor brand mention growth, referral source diversity, and consultation request quality to measure overall authority development.

Business Impact Assessment:
Connect multi-platform efforts to actual business results including lead quality, conversion rates, and client acquisition costs.

The Compound Effect of Consistent Multi-Platform Presence

Multi-platform authority building creates compound advantages that strengthen over time:

Omnipresence Effect:
Consistent presence across platforms creates the impression of market dominance and expertise, even in competitive markets.

Redundancy Protection:
Diversified platform presence protects against algorithm changes, policy shifts, or platform-specific challenges that could affect single-platform strategies.

Audience Expansion:
Different platforms attract different demographics and psychographics, expanding your potential client base beyond single-platform limitations.

Content Amplification:
Quality content shared across platforms reaches exponentially more prospects than single-platform publication.

Referral Enhancement:

Multi-platform presence makes it easier for past clients and referral sources to recommend you, as they can direct prospects to multiple touchpoints that demonstrate expertise.

The AI era has made multi-platform authority building more important than ever, as both algorithms and consumers seek evidence of expertise across multiple channels. Real estate professionals who create integrated systems with their website as the authority hub while leveraging other platforms for distribution and engagement will have significant competitive advantages in attracting and converting quality prospects.

Chapter 7

AI-GEO and Local Search Optimization for Real Estate Proessionals

Understanding the New Search Landscape

The real estate industry is experiencing a fundamental shift in how consumers discover and evaluate local expertise. Traditional search results that display lists of websites are being supplemented—and in many cases replaced—by AI-generated responses that synthesize information from multiple sources to provide direct answers.

This transformation requires real estate professionals to understand three critical new optimization strategies that go beyond traditional SEO:

AI-GEO (AI-Generated Experience Optimization):
The practice of optimizing your content and digital presence so that AI systems select your expertise when generating responses to user queries. When someone asks "What's happening in the [your city] housing market?" AI-GEO optimization helps ensure your insights are included in the AI-generated answer.

LEO (Local Experience Optimization):
The strategic optimization of your digital presence to be recognized by AI systems as a definitive local authority. This involves creating consistent local expertise signals across all digital touchpoints so AI systems understand your geographic specialization and market knowledge.

AI-LEO (Language Experience Optimization):
The optimization of your content's language and communication style to match how people naturally communicate with AI systems. This involves writing content that addresses conversational, natural language queries rather than traditional keyword-based searches. For example, optimizing for "What should I expect when selling my house for the first time?" rather than "home selling process."

These concepts represent the evolution of search optimization from competing for visibility in result lists to competing for inclusion in AI-generated expert responses that match natural human communication patterns.

How AI Systems Evaluate Local Real Estate Expertise
Modern AI systems use sophisticated algorithms to evaluate and rank local expertise when generating responses to real estate queries. Understanding these evaluation criteria is essential for positioning yourself as the recognized authority in your market.

Depth and Consistency of Local Content:
AI systems analyze the breadth and depth of your local market coverage. Real estate professionals who consistently publish detailed content about specific neighborhoods, school districts, market trends, and local developments receive higher authority scores than those with sporadic or surface-level coverage.

Geographic Signal Consistency:
AI algorithms look for consistent geographic markers across your content. When you repeatedly reference the same service areas, neighborhoods, and local landmarks across your website, social media, and other platforms, AI systems develop confi-

dence in your local expertise boundaries.

Recency and Update Frequency:
Real estate markets change rapidly, and AI systems prioritize sources that provide current information. Regular publication of market updates, new listing analyses, and changing local conditions signals active local engagement to AI algorithms.

Cross-Platform Authority Verification:
AI systems cross-reference information across platforms to verify expertise claims. When your local market insights appear consistently across your website, social media profiles, video content, and professional listings, it strengthens your authority signals.

Local Community Engagement:
AI systems can detect genuine community involvement through mentions in local publications, participation in community events, and engagement with local businesses and organizations. This engagement serves as external validation of your local expertise.

AI-LEO Strategy Development for Real Estate

Language Experience Optimization focuses on creating content that matches the conversational way people interact with AI systems, voice assistants, and chatbots. This is particularly important for real estate professionals because property decisions often involve complex, nuanced questions that consumers express in natural language.

Understanding Natural Language Query Patterns

Traditional Keyword Searches vs. AI-LEO Queries:
- Traditional: "*mortgage rates 2025*"
- AI-LEO: "*What are current mortgage rates and how do*

they affect my monthly payment?"
- Traditional: "*best neighborhoods [city name]*"
- AI-LEO: "*Which neighborhoods would be good for down-sizers who want cultural and social activities as well as easy access to health care, shopping and restaurants?*"

Question-Based Content Structure:
AI-LEO optimization requires structuring content around the actual questions people ask, complete with context and qual-ifiers. Real estate consumers rarely ask simple, keyword-based questions—they ask complex, situation-specific questions that require expert interpretation.

Conversational Content Development
Natural Language Headlines and Subheadings:
Instead of "H*ome Buying Process Overview,*" use "*What Should I Expect When Buying My First Home?*" This approach matches how people naturally ask questions to AI assistants and creates content that AI systems can easily extract and reference.

Complete Answer Architecture:
AI-LEO content should provide complete, conversational an-swers that don't require additional research. When someone asks a complex question about real estate, your content should ad-dress the question comprehensively, including context, consid-erations, and practical next steps.

Contextual Language Usage:
Include the language patterns and phrases that real estate con-sumers actually use when discussing their situations. This in-cludes emotional language, uncertainty expressions, and the way people naturally describe their real estate needs and concerns.

AI-LEO Content Examples for Real Estate

Process-Related AI-LEO Content:

- *"What happens after I make an offer on a house?"*
- *"How do I know if I'm getting a fair price in this market?"*
- *"What should I do to prepare my house for sale?"*
- *"How long will it take to find and buy a house?"*

Market and Location AI-LEO Content:

- *"Is now a good time to buy in [your area]?"*
- *"What's the real estate market like for families in [neighborhood]?"*
- *"Should I wait for prices to come down before buying?"*
- *"What makes one neighborhood better than another for investment?"*

Financial AI-LEO Content:

- *"How much house can I actually afford with my income?"*
- *"What costs should I expect beyond the down payment?"*
- *"Is it better to rent or buy in my situation?"*
- *"How do property taxes affect my monthly housing costs?"*

LEO Strategy Development for Real Estate

Local Experience Optimization requires systematic development of location-specific authority signals that AI systems can easily identify and evaluate.

GEOGRAPHIC CONTENT MAPPING

Service Area Definition:

Clearly define and consistently reference your primary service areas across all content. Use specific neighborhood names, school district boundaries, zip codes, and local landmarks that AI systems can easily categorize and associate with your expertise.

Micro-Local Expertise:
Develop deep content around specific micro-locations within your broader service area. AI systems increasingly favor hyper-local expertise over broad geographic claims. A real estate professional known as the expert on three specific neighborhoods often outranks one claiming general city-wide knowledge.

Local Entity Optimization:
Create comprehensive content around local entities that matter to real estate decisions:
- Major employers and their impact on housing demand
- Transportation infrastructure and commute patterns
- Shopping, dining, and entertainment districts
- Parks, recreation, and community amenities
- Local government and zoning considerations

Local Market Intelligence Content

Hyperlocal Market Analysis:
Develop regular content that analyzes market conditions at the neighborhood level. AI systems prioritize sources that provide granular local insights over general market commentary.

Price Trend Interpretation:
While market data is widely available, interpretation requires local expertise. Create content that explains what price trends mean for different neighborhoods and buyer segments, demonstrating analytical capability that AI systems recognize as valuable expertise.

Development Impact Assessment:
Cover new developments, infrastructure projects, and zoning changes that affect local real estate values. This type of forward-looking analysis demonstrates deep local knowledge that

AI systems struggle to find from generic sources.

Local Policy and Regulation Expertise:
Discuss local real estate regulations, tax implications, and policy changes that affect property ownership and transactions. This specialized knowledge positions you as essential for complex local real estate decisions.

AI-GEO Content Optimization Techniques

Creating content that AI systems select for inclusion in generated responses requires understanding how these systems parse and evaluate information.

QUESTION-ANSWER CONTENT STRUCTURE

Direct Question Addressing:
Structure content to directly answer common real estate questions. Use clear headings that match likely search queries: "*Is now a good time to buy in [neighborhood]?*" or "*What should I know about [school district] before buying?*"

Comprehensive Answer Development:
Provide complete answers that don't require additional research. AI systems favor sources that offer thorough responses over those requiring multiple sources to generate complete answers.

Supporting Evidence Integration:
Include specific data, examples, and local references that support your conclusions. AI systems evaluate the evidence quality when selecting sources for response generation.

STRUCTURED DATA AND FORMATTING

Schema Markup Implementation:
Use appropriate schema markup to help AI systems understand

your content context, local service areas, and expertise categories. Real estate-specific schema helps AI systems categorize your local expertise appropriately.

Clear Information Hierarchy:
Organize content with clear headings, subheadings, and logical information flow that AI systems can easily parse and extract relevant information from.

Fact and Opinion Separation:
Clearly distinguish between factual information and professional opinions. AI systems often extract facts for objective responses while crediting opinion sources for subjective guidance.

Cross-Platform LEO Implementation
Local Experience Optimization requires consistent implementation across all digital platforms to build comprehensive authority signals.

WEBSITE LEO FOUNDATION
Local Landing Page Development:
Create dedicated pages for each service area with comprehensive local information. These pages should include market data, neighborhood characteristics, local amenities, and your specific expertise in that area.

Blog Content Localization:
Ensure blog content consistently references local areas, uses local terminology, and addresses location-specific concerns. Every piece of content should reinforce your geographic expertise.

Contact Information Optimization:
Use consistent NAP (Name, Address, Phone) information

across all platforms and include local phone numbers and addresses that clearly indicate your service areas.

SOCIAL MEDIA LEO STRATEGY
Location-Tagged Content:
Consistently use location tags, local hashtags, and geographic references in social media content to reinforce local expertise signals.

Local Community Engagement:
Actively participate in local Facebook groups, Instagram location pages, and community discussions to demonstrate ongoing local involvement.

Geo-Referenced Content:
Share content about local events, market conditions, and neighborhood developments with appropriate geographic tags and references.

VIDEO PLATFORM LEO OPTIMIZATION
Location-Specific Video Content:
Create video content that showcases local areas, discusses neighborhood characteristics, and demonstrates on-location expertise.

Local SEO Video Optimization:
Use location-specific keywords in video titles, descriptions, and tags to help AI systems understand the geographic relevance of your content.

Virtual Neighborhood Tours:
Develop video content that provides virtual tours and local insights that demonstrate intimate neighborhood knowledge.

Measuring AI-GEO, LEO, and AI-LEO Success

Traditional SEO metrics don't fully capture AI-GEO, LEO, and AI-LEO performance, requiring new measurement approaches and tools.

AI-LEO Performance Indicators

Natural Language Query Performance:
Monitor how your content performs for conversational, question-based searches. Track rankings and visibility for queries that match natural speech patterns rather than traditional keyword searches.

Voice Search Optimization Results:
Measure performance for voice search queries, as these typically use more natural, conversational language patterns that align with AI-LEO optimization.

Long-Tail Conversational Keywords:
Track performance for longer, more specific queries that match how people naturally describe their real estate situations and needs.

AI Assistant Integration:
Monitor whether your content appears in responses from AI assistants like Google Assistant, Alexa, or ChatGPT when users ask real estate questions.

AI-GEO Performance Indicators

Featured Snippet Appearances:
Monitor how often your content appears in featured snippets, as these often serve as sources for AI-generated responses.

Brand Mention Tracking:

Use tools to track when your name or business appears in AI-generated responses to real estate queries, even when not directly linked.

Question-Answer Performance:

Track which of your content pieces perform well for question-based searches, as these indicate AI-GEO optimization success.

LEO PERFORMANCE METRICS
Local Query Visibility:

Monitor your visibility for location-specific real estate searches and how often your content appears in local search results.

Geographic Authority Indicators:

Track your recognition as a local expert through mentions in local media, community websites, and local business directories.

Local Engagement Metrics:

Measure engagement from users in your target geographic areas across all platforms to assess local authority building success.

The Future of AI-Powered Local Search

Understanding current trends in AI development helps prepare for the continued evolution of local search and expertise evaluation.

Voice Search Integration:

As voice search becomes more prevalent, AI systems will increasingly need to identify trusted local sources for spoken responses to real estate questions.

Personalized Local Recommendations:
AI systems are developing more sophisticated personalization capabilities, making consistent local expertise demonstration crucial for inclusion in personalized recommendations.

Real-Time Local Intelligence:
Future AI systems will likely integrate real-time local data, making current market knowledge and active local engagement even more valuable for authority building.

Cross-Platform Intelligence Integration:
AI systems continue improving their ability to synthesize information across platforms, making consistent multi-platform local expertise demonstration essential.

Implementation Roadmap for AI-GEO, LEO, and AI-LEO

Successful implementation of AI-GEO, LEO, and AI-LEO strategies requires systematic approach and consistent execution over time.

Phase 1: Foundation Building (Months 1-3)
- Audit current content for local references and AI-optimization opportunities
- Implement schema markup and structured data across your website
- Begin rewriting key content using natural language, conversational approaches (AI-LEO)
- Optimize existing content for question-answer format
- Start incorporating conversational query patterns into new content

Phase 2: Authority Development (Months 4-9)

- Expand local content coverage across all service areas (LEO)
- Increase cross-platform local expertise demonstration
- Develop comprehensive FAQ sections using natural language questions (AI-LEO)
- Build local community engagement and participation
- Create content that addresses complete conversational queries rather than keyword fragments

Phase 3: Optimization and Scaling (Months 10-12)

- Refine content based on AI-GEO, LEO, and AI-LEO performance data
- Expand into emerging local topics and market developments
- Build strategic local partnerships and collaborations
- Establish thought leadership in local real estate trends
- Optimize for voice search and AI assistant queries using natural language patterns

The integration of AI into search and local discovery represents both a challenge and an unprecedented opportunity for real estate professionals. Real estate professionals who understand and implement AI-GEO and LEO strategies will establish sustainable competitive advantages that become more valuable as AI systems become more sophisticated and prevalent in consumer research patterns.

Chapter 8

Integration Strategy - Your Website as Authority Hub

Creating a Unified Digital Authority System

Throughout this book, we've explored why real estate professionals need their own websites, how to build authority through content, and how to optimize for AI-powered search systems. Now comes the crucial step: integrating all these elements with your existing real estate tools and broker resources to create a comprehensive business development system.

The goal isn't to replace your current tools but to position your website as the central hub that coordinates and amplifies all your marketing efforts while building long-term digital equity in your personal brand.

The Hub and Spoke Integration Model

Your website serves as the authoritative hub while other platforms and tools function as specialized spokes that drive traffic, engagement, and leads back to your owned platform.

Website Hub Functions:
- Comprehensive content library demonstrating complete expertise
- Lead capture systems with valuable resource offerings
- Email list building and nurture system management
- SEO and AI-GEO optimization for long-term organic traffic
- Professional presentation and trust signal optimization
- Analytics and performance measurement centralization

Spoke Platform Integration:

- **IDX and MLS Tools**: Complement rather than compete with broker resources
- **Social Media Platforms:** Drive traffic to comprehensive website content
- **Email Marketing**: Nurture relationships while directing to website resources
- **Video Platforms**: Showcase expertise while promoting detailed written content
- **CRM System**s: Track website-generated leads alongside traditional sources
- **Local Directories**: Reinforce local authority while directing to website

IDX and Broker Resource Integration

One of the most common concerns real estate professionals have about building their own website is how it will work alongside their broker's IDX system and existing tools. The solution lies in strategic complementarity rather than competition.

Complementary Positioning Strategy:

Your broker's IDX system excels at property search functionality and MLS integration. Your website excels at expertise demonstration and relationship building. Position your website as the place prospects go to understand market conditions, neighborhoods, and the buying/selling process, while using your broker's IDX for actual property search.

Cross-Platform Linking:

Include links to your broker's IDX system when appropriate, particularly in market analysis content where you discuss specific properties or market segments. This shows collaboration rather than competition while directing qualified prospects to property search tools.

Lead Attribution Systems:
Implement tracking systems that help you understand which leads originate from your content marketing efforts versus IDX or other sources. This data helps demonstrate the value of your authority building efforts to broker management.

Educational Content Integration:
Use your website to educate prospects about how to effectively use IDX systems, interpret property data, and understand market information. This positions you as the expert guide who helps clients navigate available tools successfully.

CRM and Lead Management Integration
Your website authority building efforts should integrate seamlessly with your existing CRM and lead management systems to create comprehensive prospect development workflows.

Lead Source Tracking:
Tag website-generated leads with specific source information to track which content pieces and lead magnets produce the highest-quality prospects. This data informs content strategy and helps demonstrate ROI.

Nurture Sequence Coordination:
Coordinate email nurture sequences from your website with CRM-based follow-up systems. Website subscribers who download resources should enter appropriate CRM workflows based on their indicated interests and timeline.

Content Consumption Tracking:
Monitor which prospects consume specific content pieces to in-

form sales conversations and provide personalized consultation approaches based on demonstrated interests.

Referral Source Enhancement:

Use website content to support referral relationships by providing partners with specific resources they can share with potential referrals, making their recommendations more valuable and specific.

Social Media Traffic Direction Strategy

Rather than treating social media as a separate marketing channel, integrate it as a traffic direction system that drives engaged prospects to your comprehensive website content.

Content Teaser Strategy:

Share highlights, insights, or introductory information on social platforms while directing followers to your website for complete analysis and detailed resources.

Platform-Specific Adaptation:

Adapt content for each platform's format and audience expectations while maintaining consistent messaging that drives traffic to your website for comprehensive information.

Engagement to Conversion Funnels:

Create clear pathways from social media engagement to website visits to email subscriptions to eventual consultation bookings.

Cross-Platform Content Promotion:

Use social media to promote new website content, lead magnets, and resources while building anticipation for comprehensive analysis and expert insights.

Email Marketing Ecosystem Development

Email marketing serves as the crucial bridge between all your marketing efforts, providing direct communication with your most engaged prospects while supporting overall authority building goals.

List Building Integration:

Every piece of content, social media post, and marketing effort should include opportunities for email list building through valuable resource offerings and expert insight subscriptions.

Content Distribution Coordination:

Use email to notify subscribers about new website content, up-coming market analysis, and exclusive insights while driving traffic back to your website for detailed information.

Segmentation and Personalization:

Segment email lists based on content consumption patterns, geographic interests, and engagement levels to provide personalized communication that supports individual prospect development.

Multi-Touch Attribution:

Track how email subscribers interact with website content, social media, and other touchpoints to understand complete customer journey patterns and optimize integration strategies.

Local Community Integration

Your digital authority building efforts should complement and enhance your local community involvement rather than replacing face-to-face relationship building.

Event Content Creation:

Document community involvement, local market presentations,

and professional speaking engagements to create content that demonstrates local engagement and expertise.

Local Partnership Amplification:
Use website content to highlight relationships with other local professionals, community organizations, and business partnerships that reinforce your local authority.

Community Resource Development:
Create website resources that serve the broader community, such as local vendor directories, community event calendars, and neighborhood information guides that position you as a community resource.

Local Media Integration:
Leverage local media appearances, expert commentary opportunities, and community recognition through website content that amplifies your local authority building efforts.

Analytics and Performance Integration

Effective integration requires comprehensive analytics that track performance across all marketing channels while attributing business results to specific efforts.

Multi-Channel Attribution:
Implement tracking systems that help you understand how prospects interact with content across platforms before converting to leads and eventually clients.

Content Performance Analysis:
Monitor which content pieces generate the most engagement, lead conversions, and eventual business to inform future content strategy and resource allocation.

ROI Measurement:
Track the business impact of authority building efforts by measuring lead quality, conversion rates, and client acquisition costs for website-generated prospects versus other sources.

Optimization Feedback Loops:
Use performance data to continuously refine integration strategies, content approaches, and cross-platform coordination efforts.

Long-Term Business Development Strategy

The integration of website authority building with existing tools creates compound advantages that strengthen over time and build genuine business equity.

Digital Asset Development:
Every piece of content, email subscriber, and authority signal builds equity in your personal brand that transfers with you regardless of brokerage changes or market conditions.

Referral Network Enhancement:
Website content and resources make referral partners more effective by providing specific, valuable materials they can share when making recommendations.

Market Position Strengthening:
Consistent authority demonstration across integrated platforms creates increasingly difficult competitive barriers as your expertise library and local recognition grow.

Business Transition Preparation:
Whether planning retirement, brokerage changes, or business succession, owned digital assets provide transferable value that

extends beyond immediate transaction income.

Implementation Timeline and Milestones
Successful integration requires systematic implementation over time rather than attempting to activate all systems simultaneously.

Phase 1: Foundation (Months 1-3)
- Establish website with basic content and lead capture systems
- Integrate with existing CRM for lead tracking and attribution
- Begin basic social media traffic direction strategies
- Implement fundamental analytics and tracking systems

Phase 2: Content Authority Building (Months 4-9)
- Expand content library with regular publication schedule
- Develop comprehensive lead magnets and nurture sequences
- Increase cross-platform integration and traffic direction
- Build local community content and partnership integration

Phase 3: Optimization and Scaling (Months 10-18)
- Refine integration based on performance data and business results
- Expand into advanced AI-GEO and LEO optimization strategies
- Develop sophisticated attribution and ROI measurement systems
- Scale successful integration strategies across all marketing efforts

Overcoming Common Integration Challenges

Most real estate professionals encounter predictable challenges when integrating website authority building with existing systems. Understanding these challenges and their solutions helps ensure successful implementation.

Time Management Concerns:
Start with systematic content creation and repurposing strategies that maximize efficiency. One comprehensive piece of content can provide material for multiple platforms and marketing touchpoints.

Technology Integration Complexity:
Begin with basic integration approaches and gradually add sophistication as comfort and understanding increase. Most integration benefits come from strategic coordination rather than complex technical implementation.

Broker Relationship Management:
Position website authority building as enhancement and support for broker resources rather than competition. Share traffic and lead attribution data that demonstrates value addition rather than replacement.

ROI Justification:
Track and report on lead quality, conversion rates, and business impact to demonstrate the value of authority building efforts to yourself and stakeholders.

The Compound Effect of Strategic Integration

When properly implemented, the integration of website authority building with existing real estate tools creates compound advantages that strengthen over time:

Authority Amplification:

Each marketing touchpoint reinforces expertise demonstration, creating stronger overall authority signals than any single platform could achieve.

Efficiency Multiplication:

Strategic integration allows single content creation efforts to serve multiple marketing functions, maximizing time investment returns.

Lead Quality Enhancement:

Prospects who interact with multiple touchpoints before converting typically become higher-quality clients who appreciate expertise and provide better referrals.

Competitive Differentiation:

Integrated authority building creates sustainable competitive advantages that become more valuable as more real estate professionalss recognize the importance of digital expertise demonstration.

The integration of website authority building with existing real estate tools represents a strategic evolution rather than a marketing revolution. By positioning your website as the hub that coordinates and amplifies all other efforts, you create a sustainable competitive advantage while building genuine business equity that appreciates over time.

INDEX

Appendix A: Quick Reference Guides

A.1: AI-GEO Optimization Checklist

- ☐ Content structured for question-answer format
- ☐ Clear headings that match search queries
- ☐ Comprehensive answers that don't require research
- ☐ Supporting evidence and local examples included
- ☐ Schema markup implemented
- ☐ Cross-referenced internal linking

A.2: LEO Implementation Checklist

- ☐ Consistent NAP (Name, Address, Phone)
- ☐ Location-specific landing pages created
- ☐ Local keywords integrated naturally
- ☐ Geographic schema markup added
- ☐ Google My Business optimization completed
- ☐ Local citation building initiated

A.3: AI-LEO Content Optimization

- ☐ Headlines written as natural questions
- ☐ Content addresses conversational queries
- ☐ Natural language patterns used throughout
- ☐ Voice search queries considered
- ☐ Long-tail conversational keywords integrated
- ☐ FAQ sections using natural speech patterns

A.4: Trust Signal Audit

- ☐ Professional credentials displayed prominently
- ☐ Client testimonials with specific results
- ☐ Local market expertise demonstrated
- ☐ Professional photography used
- ☐ Contact information easily accessible
- ☐ Social proof elements integrated

Appendix B: Content Templates and Frameworks

B.1: BLOG POST TEMPLATES
Market Analysis Template:
- ☐ **Headline**: "*What [Current Event] Means for [Local Area] Real Estate*"
- ☐ **Opening**: Current market data/statistics
- ☐ **Analysis**: Local implications and interpretation
- ☐ **Buyer/Seller Impact**: Specific advice for each audience
- ☐ **Call-to-Action**: Link to consultation or resource

Neighborhood Guide Template:
- ☐ Introduction: *Why this neighborhood matters*
- ☐ Demographics and lifestyle fit
- ☐ Transportation and commuting
- ☐ Local amenities and entertainment
- ☐ Market trends and price ranges
- ☐ Investment potential assessment

B.2: LEAD MAGNET TEMPLATES
Buyer's Guide Framework:
- ☐ Cover page with professional branding
- ☐ Table of contents
- ☐ Introduction to local market
- ☐ Step-by-step buying process
- ☐ Neighborhood comparisons
- ☐ Financial considerations
- ☐ Timeline expectations
- ☐ Local resources and contacts

Market Report Template:
- ☐ Executive summary
- ☐ Key statistics and trends
- ☐ Price analysis by area
- ☐ Inventory levels and implications
- ☐ Expert predictions and recommendations

Appendix C: Implementation Worksheets

C.1: DIGITAL AUTHORITY AUDIT WORKSHEET
Current Platform Assessment:
- ☐ List all current digital platforms used
- ☐ Evaluate content ownership on each platform
- ☐ Assess traffic and engagement sources
- ☐ Identify platform dependency risks
- ☐ Calculate potential loss if platforms changed

Content Inventory:
- ☐ Catalog existing valuable content
- ☐ Identify repurposing opportunities
- ☐ Assess content gaps in local expertise
- ☐ Plan content migration to owned platform

C.2: LOCAL MARKET CONTENT PLANNING
Service Area Definition:
- ☐ Primary neighborhoods served
- ☐ Geographic boundaries
- ☐ Demographic characteristics
- ☐ Local market specializations

Content Opportunity Mapping:
- ☐ Recurring market topics (monthly reports)
- ☐ Seasonal content opportunities
- ☐ Local event and development coverage
- ☐ Educational content gaps
- ☐ Competitive content analysis

C.3: LEAD MAGNET DEVELOPMENT WORKSHEET
Audience Problem Identification:
- ☐ Common buyer questions and concerns
- ☐ Seller challenges and pain points
- ☐ First-time buyer education needs
- ☐ Investment buyer information gaps
- ☐ Local market confusion areas

Resource Development Planning:

- ☐ Content format selection (PDF, video, tool)
- ☐ Information gathering requirements
- ☐ Design and production timeline
- ☐ Distribution and promotion strategy
- ☐ Performance measurement plan

Appendix D: Measurement & Analytics

D.1: Key Performance Indicators (KPI's)

Website Authority Metrics:

- Organic traffic growth
- Time on site and page views
- Email subscription conversion rates
- Content engagement metrics
- Local search ranking improvements

Business Impact Metrics:

- Lead source attribution
- Lead quality scoring
- Consultation booking rates
- Client conversion percentages
- Average transaction values

D.2: Analytics Setup Guide

Essential Tracking Implementation:

- Google Analytics 4 configuration
- Google Search Console setup
- Email marketing platform integration
- Social media analytics connection
- Lead source tracking systems

D.3: ROI Calculation Frameworks

Content Marketing ROI:

- Content creation time investment
- Lead generation attribution
- Client acquisition cost calculation
- Lifetime value consideration
- Competitive advantage valuation

Appendix E: Legal and Compliance Considerations

E.1: Privacy and Data Protection

- □ GDPR compliance for email capture
- □ State privacy law requirements
- □ Data retention and deletion policies
- □ Cookie consent implementation
- □ Third-party integration privacy

E.2: Real Estate Compliance:

- □ MLS rule compliance for content
- □ Fair housing considerations in content
- □ Advertising regulation adherence
- □ Professional licensing requirements
- □ Broker policy alignment

E.3: Content Legal Guidelines

- □ Copyright considerations for images
- □ Attribution requirements for data
- □ Testimonial and review guidelines
- □ Social media compliance rules
- □ Professional photography contracts

Appendix F: Resources and Tools

F.1: Recommended Tools and Platforms

Website Development:
- □ WordPress hosting recommendations
- □ Theme suggestions for real estate
- □ Essential plugin lists
- □ Design tool recommendations

Content Creation:
- □ Writing and editing tools
- □ Image creation and editing software
- □ Video production resources
- □ Content calendar applications

Analytics and Measurement:
- □ Free analytics tools
- □ Paid monitoring services
- □ Social media management platforms
- □ Email marketing service comparison

F.2: Industry Resources
- □ National Association of Realtors® research
- □ Local MLS data sources
- □ Market analysis tools
- □ Professional development opportunities

F.3: Further Reading
- □ Digital marketing strategy books
- □ Real estate industry publications
- □ AI and search engine optimization resources
- □ Business development and authority building materials

NOTES

NOTES

NOTES

NOTES

NOTES

NOTES

www.ingramcontent.com/pod-product-compliance
Lightning Source LLC
Chambersburg PA
CBHW071606200326
41519CB00021BB/6891